The Evolution of Planing Sailboats

a monograph

by John MacBeath Watkins

The Evolution of planing sailboats

by John MacBeath Watkins

A friend in England announced that he had bought a boat, an International 14 designed and built by Uffa Fox, the man who, he said, invented the planing dinghy.

I told him I was a great admirer of Uffa Fox, but had in my possession lines for a planing dinghy designed by Nathaniel Herreshoff in 1900. This upset him, and led to a thread about the history of planing boats on the WoodenBoat forum, which disappeared, and then another thread on the same forum on the same subject, which unearthed a lot more information in a collaborative bit of research involving people in England, New Zealand, Australia, Canada, the United States, and probably some countries I'm neglecting to mention. Contributors included two designers, a model builder for a test tank, a maritime museum curator and a number of experienced sailors. Special thanks to Richard Woods of Woods Designs and Ben Fuller, curator of the Penobscot Marine Museum. This is a narrative of what we learned.

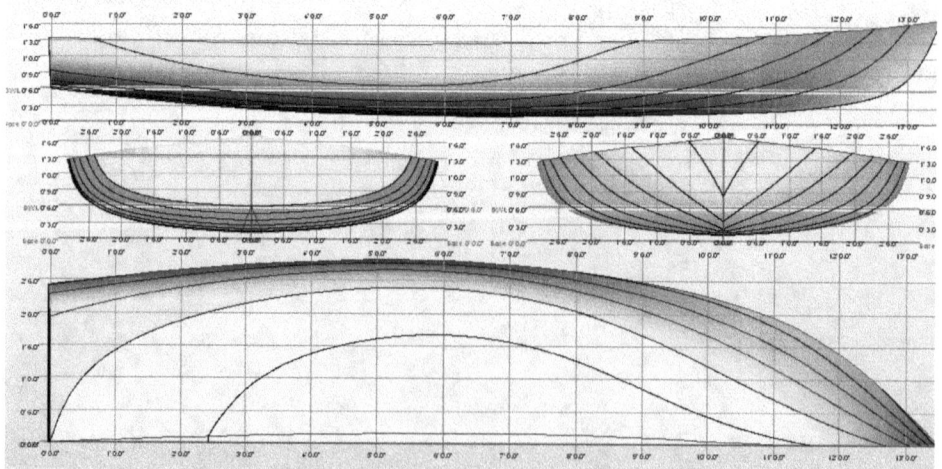

A representation of the type of planing sailboat Nathaniel Herreshoff was designing in 1900.
Illustration by John MacBeath Watkins

I don't have the rights to the lines Howard Chapelle drew up of Herreshoff's 13 ½ foot spritsail boat, so I've drawn up a set of lines in Delftship that represent the shape as closely as I can working by eye. I've made no measurements, and don't claim this is extremely accurate, but it will give you a good idea of the boat under discussion. She has a broad bow, because she was a rule-beater. The class she raced with, spritsail boats based on and including the Woods Hole workboats, required that she be no more than 13½ feet long, and carry a spritsail with no boom. So, to give the boat the longest foot and the largest sail possible, Herreshoff mounted the mast on top of the stem. It was mounted in a bronze tube, supported by flying buttresses that went back and sideways to the deck. The boat is similar to a sneakbox, a type known for its speed off the wind, but the bow is sharper.

Planing sailboats were not necessarily the fastest type for all uses, although most of the short-distance speed records for sailboats since the last multihull record in 1980 have been set by planing "vessels," if you can apply that to windsurfers, kiteboards, and *Vestas Sailrocket 3*, which rides on three stepped hulls and uses foils.

The first challenge to displacement monohulls in organized sailboat racing came in 1876, when Nathaniel Herreshoff designed a racing catamaran that blew past the fastest racing boats of her size at the time, sandbagger sloops, and catamarans were quickly banned from racing against monohulls. Herreshoff found himself racing the fastest steamers instead. A few other builders took up building racing catamarans, but the type eventually pretty much died out until after World War II.

When planing monohulls were introduced, they were not so easily banned, so for many years, they were the fastest boats allowed to race in most fleets. Even now, when there are plenty of one-design catamarans around with fleets to race in, people persist in sailing in planing dinghies, because they like the way they sail.

Sailing is a sensual sport. Going 15 mph in my car feels slow, in an airplane I'd never even leave the ground, but when a 15-foot dinghy is sailing that fast with no noises but the splash of the thrown spray and the whooping of the crew, pushed through the water by the quiet power of the wind, it feels glorious. Catamarans sail fastest with the weather hull out of the water, and the crew farther from the spray. It's a different sensation, on a faster boat. But within a racing fleet, what's important is not your top potential speed, but your speed relative to the other boats in your class. And there are plenty of one-design fleets of planing dinghies, because the sensation of sailing them is pleasurable.

Planing makes it possible for a boat to move at higher speed with less power, because hydrodynamic lift reduces the immersed volume and the water breaks from the hull cleanly, freeing the boat from its wave pattern.

Because work boats are usually designed to carry loads, most sailing work boats are not designed to plane. Some will catch a wave and surf; and there are records of Norwegian vessels completing runs faster than hull speed.

The book *Inshore Craft of Norway,* by Bernhard Faeroyvik and Oystein Faeroyvik, mentions a day's trip in a holmedalsjekta (30-35 ft.) with a cargo of 11 head of cattle. Jens Sorensen Bakke, born 1866, told of a memorable voyage as a boy, sailing from Sunnfjord to Bergen, leaving at 9 a.m. and arriving at 4 p.m., having delivered the cattle along the way. "The distance was 126 km, which means an average speed of 9.7 knots," the authors dryly inform us. The hull speed of such a boat would be about 8 to 8.3 knots. We don't know how long it took to stop and unload the cattle, so I'd say at a minimum, the boat was spending a lot of time at semi-displacement speed, if not actually surfing much of the way. This would be a square-sterned boat with a cabin in the stern, much like the one in the picture on the cover of the book recounting that remarkable voyage.

Another of the few commercial cases of using hydrodynamic lift prior to powerboats were express canal boats used in Scotland and parts of England. They did so by a process similar to planing, but not the same thing.

Around 1830, a canal boat proprietor named William Houston had an amazing experience when the horse towing his vessel took fright and set off in a panic, while Houston hung on, thinking the drag of the boat would tire the horse. Instead, the boat rose up on its bow wave and went forward at great speed, while not seeming to tire the horse much. The stern wave most vessels would produce at high speed had disappeared.

Houston perceived a commercial possibility, and determined that in the shallow water of the canal, a light, flat-bottomed vessel would rise up and proceed at about 12 mph with much less than the usual amount of drag. The usual speed had been 4 mph.

In 1835, Houston's express canal boats were making an eight-mile run at 10 mph, and carried 323,290 passengers in that year alone. By 1840, railroads were already taking away the business, but that year a scientist named John Scott Russell published a scientific study of the express canal boats.

Russell showed that the speed these boats (called fly-boats in England) achieved was related to the depth of the water. A speed of 10-12 mph is not planing speed for a boat the length of these vessels, about 65 ft. In deeper water, the boat would not achieve sufficient hydrodynamic lift to climb out of the "hole" of its bow and stern wave until it reached about twice that speed. (The formula is 2.4 times the square root of the waterline length in feet = planing speed in knots. Hull speed is 1.35 times the square root of the waterline length, and at the speeds in between, the boat is described as semi-planing or semi-displacement.) The canal boats were a special case where the main benefit of planing – achieving enough speed to reduce the vessel's immersed volume – could occur at a lower speed than normal. Today, most attempts to set new records for speed under sail on the water are done in the shallowest water possible. But the word "hydroplane," from which we derive "plane" in the sense of taking advantage of hydrodynamic lift to go fast, was coined in 1904 and attested as a verb in 1914, so much of our early history of this phenomenon was not described as planing. This creates a problem with language. For example, prior to 1904 even planing powerboats were not described as planing, and several histories of powerboats have therefore treated vessels prior to the invention of the language of planing as displacement boats.

Consider the racing powerboat *Standard*, 60 ft. long and capable of 30 mph.

To plane, a boat needs to be going at least 2.4 x the square root of the waterline length,

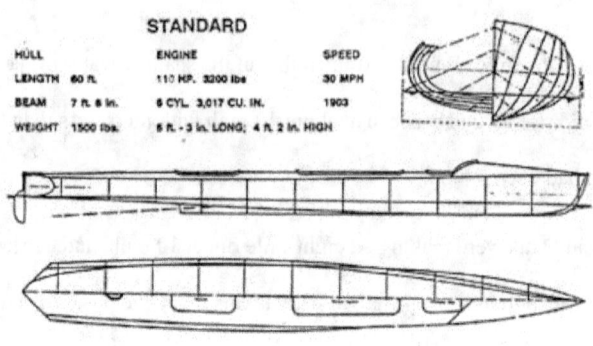

STANDARD

HULL		ENGINE		SPEED	
LENGTH	60 ft.	110 HP. 3200 lbs		30 MPH	
BEAM	7 ft. 6 in.	6 CYL. 3,017 CU. IN.		1903	
WEIGHT	1500 lbs.	5 ft. - 3 in. LONG; 4 ft. 2 in. HIGH			

expressed in nautical miles per hour (knots.) 30 mph is about 26 knots, and the square root of 60 is about 7.75, so *Standard* was well into the planing speed range. The boat achieved this with a long, flat run that provided plenty of lift, yet the histories written of powerboats typically class this as a displacement hull, because contemporary writers did not say it was planing. Of course, they didn't say it was planing because the term wasn't yet invented, but they did record the vessel's speed, and if you do the math, you must conclude that she was planing, or else defying the laws of hydrodynamics.

Powerboats can plane based on applying lots of power to a hull of the right shape, even if the boat is fairly heavy. Sailboats have a tougher time developing high horsepower, and therefore must typically be lighter. George O'Day wrote an article, *So you Want to Plane,* in Yachting, Vol. 101, No.4, 1957, pg.82, which essentially says that in addition to hull shape criteria, you need a displacement/length ratio in the 80-125 range, a sail area/displacement ratio of about 28 or more, and a sail area/wetted surface ratio of at least 2.5. You also need to be able to give the boat the proper sail shape and develop plenty of power to carry sail. Of course, O'Day was not saying that a boat lighter than a displacement/length ratio of 80 won't plane, he was saying the some boats, better shaped and with more power to carry a big rig, will plane even at a displacement/length ratio of 125, while other shapes and rigs need to be as light as 80 or less.

For a hull to plane, it needs a reasonably flat surface. How flat the hull is fore and aft is called the rocker, how flat it is in section is called the deflection angle. A deep V hull has a greater deflection angle than a shallow V hull.

Think of deflection angle this way; the water supporting the hull is presumably pushing straight up on the bottom, and the bottom, as the boat moves, is pushing the water away at the angle of the deadrise. This is most simply illustrated with a V hull. While a flat-bottomed hull, when upright, has a deflection angle of zero, with a V-bottomed hull, the deflection angle is the angle of deadrise, that is, the angle at which the V meets the water.

As a hull travels forward, some of the water is pushed forward, some is pushed backward, and

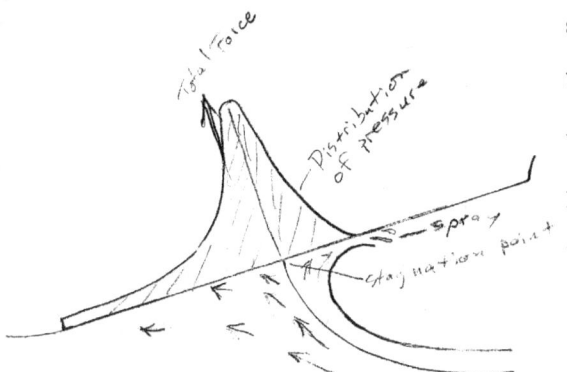

some is pushed to the sides. The point between the water going forward and the water moving aft is called the stagnation point, and is where the greatest pressure on the hull is.

Of course, a table isn't the right shape for seaworthiness, so most boats have some rocker and either some V or some U shape in section. These are compromises to make the boat safer, but the perfectly flat table is a very good shape for planing.

In most workboats, getting home without capsizing the boat is the most important thing, far more important than speed, and most are designed to carry heavy loads. The planing sailboat, therefore, is more likely to be used for pleasure than for work. And as we shall see, most of the early planing boats were either workboats modified for use in racing, or boats deliberately designed for the sport of racing.

Racing was more dangerous than not racing, but on the other hand, it was and is safer to race sailboats than almost anything else people race. Unlike fast cars and aircraft, the physics are still in your favor if you crash, and the cost of going fast this way can be minimal. Sailing at any time is a challenge, requiring the sort of skill most people won't bother to achieve, and a pleasure as you make your bargain with nature to extract the power to progress without disturbing its peace and balance. But it is in the nature of human endeavor and human play that we like to do things well, and test ourselves against each other to find out how well we're doing.

The result, in the case of sailboats, is sailboat racing. Professional mariners often had reason to wish to go fast, to chase or escape in the case of piracy, smuggling, or war, or to reach port before your cargo spoiled in the case of the fruit trade, or to reach the market before anyone else and get the best price, or to move information as quickly as possible.

The first boats called "yachts" were boats of this sort, small vessels to carry a few people and messages quickly. The word "yacht" is Dutch, and they were the first nation to widely adopt sailing for pleasure, using, of course, yachts. And below the size of the yachts were boats, often ship's boats or fishing boats for the poorer fishermen, and eventually, people started sailing these for pleasure as well. Much of the history of the sailing yacht occurs in a boating culture more ubiquitous and less prestigious than the world we think of as yachting. Small boats brought the pleasure of sailing to blue-collar workers.

Soon enough, they were testing their mettle against each other, and trying to build faster boats than each other. The evolution of sailing boats occurred in a sort of punctuated equilibrium, each breakthrough followed by a period of stability in which boats evolved slowly, piling up refinements. This is the story of one of a breakthroughs, the planing sailboat, and of the breakthroughs within that stream of evolution.

It isn't always easy to trace the history of small boats. Their existence tends to be acknowledged by historians, but few detains are recorded. Most were built by eye, and unlike larger ships where sometimes a model was made so that the authorities could understand what they were buying, seldom

was the shape recorded.

However, as people began to enjoy boats more, they began to notice them more, and record them in art. On this page is a sample from a collection of etchings done by E. W. Cooke in the 1820s and '30s.

To sail to windward, they needed a fin that would prevent the boat from sliding down wind, and for most of the history of small sailing boat in European culture, this was managed by building the boat down to a straight keel running from the stempost to the sternpost. There were exceptions. In Central America and China, daggerboards were in common use, and the Dutch have used leeboards since the 1500s. But these were not commonly used on fast boats.

In 1811, the three Swain brothers of New Jersey patented a new development for getting shallow boats to sail to windward, the centerboard. Like the leeboard, it was pivoted at the forward end and would kick up when a boat hit a reef, and like a daggerboard, it was housed in a case, usually set amidships. The first recorded centerboards appeared in Hudson River sloops, which would have been about 60 feet long, in 1815 or '16.

The centerboard has a major advantage over the daggerboard for use in shallow water, where it will kick up instead of suddenly stopping the boat as a daggerboard does when you run aground. While American and European fishermen had shown little enthusiasm for daggerboards, they quickly adopted the centerboard for a variety of types.

Some of the most graceful and quick small boats have evolved for the pursuit of shellfish, which tend to reside in shallow water and should be brought to market before they spoil. Soon after the cargo sloops started using the new invention, oyster fishermen began using centerboard boats in their trade in the New York area. By the 1850s, shallow, beamy, and surprisingly fast and nimble oyster boats were common in coastal New York, especially in Great South Bay.

Bob Fish of New York and later New Jersey was one of the men building these boats. He built the 16-foot *Una* in about 1851, and by 1852 she'd been imported to Britain, where she became known for her speed and nimble handling at the yachting center of Cowes. Dixon Kemp, author of *A Manual of Yacht and Boat Sailing,* published in several editions starting in 1878, published lines and offsets taken from the lift model Fish used to design the boat, as well as her specifications and the measurements of her rig. He also published in the same book lines for an 1870 "Una boat" designed and built in Britain, explaining that centerboard boats had gained popularity as a result of the introduction of Una. The British soon showed a preference for the sloop rig, but continued racing centerboard boats, eventually forming the 14-foot class that would be revolutionized by the planing dinghy.

But long before that happened, racing in small centerboard boats became common in America. It started among the oyster fishermen, who moored their boats in shallow water the bigger vessels didn't use, and whose haunts soon had pubs on their shores. Most of the early builders of fast centerboard boats in America were oyster fishermen or publicans. They used first bags of oysters, then bags of sand or gravel, as ballast to get the boats to go faster to windward, and the boats came to be called sandbaggers.

Contestants often bet on the races, and there were often fights over collecting the bets. Soon, rich men were buying fast boats and betting large sums on the races, just as they did with horses. For example, Nicholas Duryea, owner of the famous sandbagger Susie S, rigged a race, claimed the stakes, ($2,000, a lot of money in the 1870s) and was shot by a business partner before his opponent could recover his money in a lawsuit.

This was not gentlemanly, and hiring a bunch of ruffians to move more than 1,000 lb. of sandbags on each tack was expensive, and not convenient. Moreover, the shape of a typical sandbagger was not conducive to planing – reproductions of these historic boats reach hull speed quickly and do not typically exceed it. Their speed is restricted by the length of their wave system, which is dependent on the length of the boat.

To see why this is the case, consider our discussion of deflection angle in relation to boats built around a long, straight keel.

If you build around a full-length keel, and build down to it fore and aft, the forward and after parts of the boat are going to be deeply veed, giving the boat a deflection angle that pushes the water to the sides rather than lifting the hull.

Consider the lines John Hysolp took off of A. Cary Smith's famous sandbagger, *Comet*.

Ben Fuller, curator of the Penobscot Maritime Museum, reports that he tried towing a replica of Comet behind a powerboat, and could not get it to plane. Partly, this may be attributable to the weight of the boat, especially as raced, when even such a small boat may have required shifting 1,000 lb. or

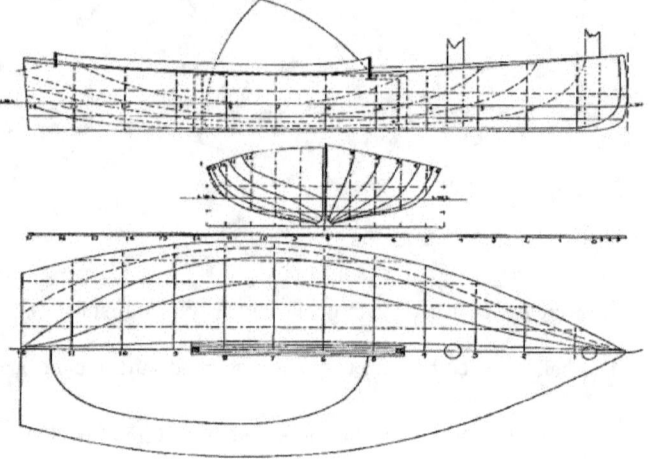

Here is Cary Smith's Comet *of 1862 as presented in the work of W. P. Stephens. Although a close friend of the designer, Stephens had a different memory of* Comet's *dimensions than did Cary Smith himself. The boat raced successfully under both cat and sloop rigs.*

more of sandbags on every tack. But part of the problem was that as long as the boats were built down to a full-length keel, they were unlikely to evolve into planing boats, and New York oystermen stuck with the full-length keel even as the boats evolved into some of the fastest racing boats of their time.

But they were not the only oystermen interested in racing.

In the Chesapeake, log canoes were used for oystering, and they were rigged for sail with a rig similar to the New Haven oyster sharpies, flat-bottomed boats that carried the leg-of-mutton sprit rig, a triangular sail with a sprit boom going from some distance above the tack across the sail to the clew. This made the sail self-vanging, and particularly fast on a reach. The log canoe sailors liked sport, so they developed a racing rig for their boats using the leg of mutton sprit rig with a large club at the clew to increase sail area, and a balanced jib as well. The rig shows clearly on the cover of a book I'd like to recommend, *Chesapeake Bay Log Canoes and Bugeyes*.

The racing log canoes will not plane, according to a friend named T.C. Price, who sails them. However, the same rig was adopted as the racing rig for the New Haven sharpies, which are flat bottomed and have an easy rocker. The speeds reported for these boats make it clear they were exceeding hull speed regularly, and by a big enough margin that they must have been planing.

Consider the Betsy D, 35 feet long and about 7 feet wide at the rub rail, completely flat bottomed, about the size of a racing log canoe, but carrying only a little more than 400 square feet of sail. She is a reproduction of the sort of working oyster sharpie used in the 1870s and 1880s. She'd have been worked by two men, and filled every day to the top of the centerboard case and the comings.

But for racing, these boats could carry a different rig, the same rig the log canoes carried.

C.P. Kunhardt, a cutter crank who actually had little use for sharpies, reports that the 35-foot Carrie V, in racing trim, carried 30 yards of sail in the jib, or 270 square feet, 75 yards (675 sq. ft.) in the foresail, 60 yards (540 sq. ft.) in the aft sail, which he calls the main and I call the mizzen, 45 yards (405 sq. ft.) in the staysail, which can only be set reaching, and 40 (360 sq. ft.) in the square sail, also only set off the wind. That's about 2,250 square feet of sail on a boat whose hull weighed about 2,000 to 2,500 lb., according to Kunhardt.

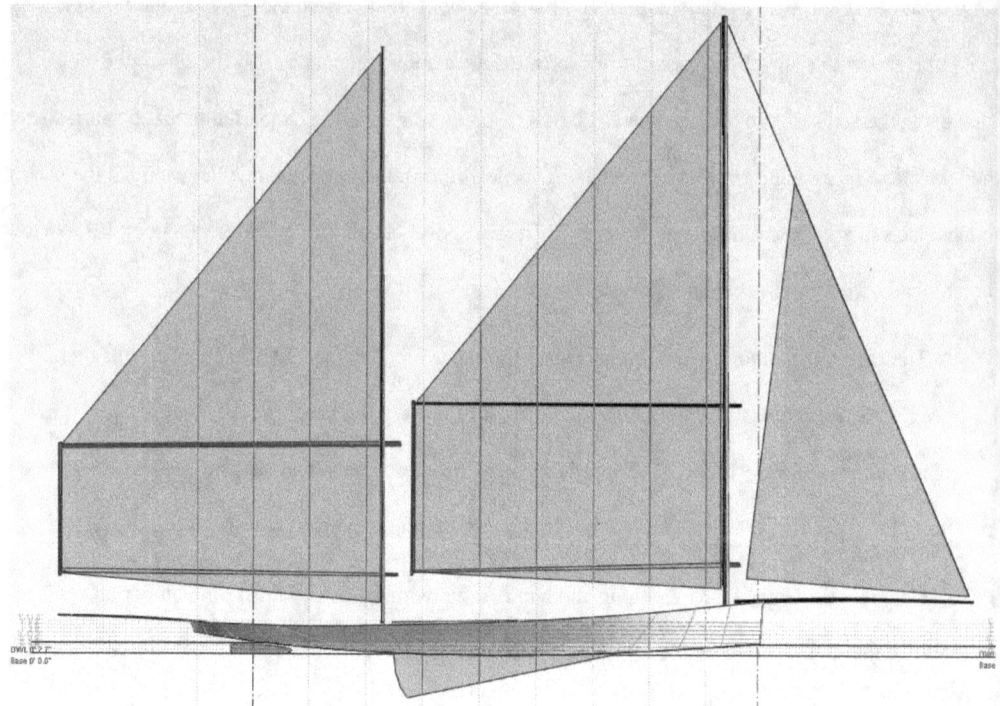

A New Haven racing sharpie with a racing rig that would carry about 1,000 square feet to windward, on a hull that would have displaced less than two tons fully crewed. Illustration by John MacBeath Watkins

The power to carry sail on these slender, shallow hulls was produced with no dead weight of ballast. Instead, the working crew of two was augmented with ten more men, nine of them assigned to the two 16-foot springboards, which were braced under the lee deck and extended about 9 feet to windward. Each springboard had a "captain," whose job it was to keep the weather chine just kissing the water. This would allow the lee chine to dig in and help the boat go to windward, and given them a low deflection angle so that they could generate plenty of hydrodynamic lift. The hull and crew together would have weighed about 4,000 lb. on a 35-foot hull.

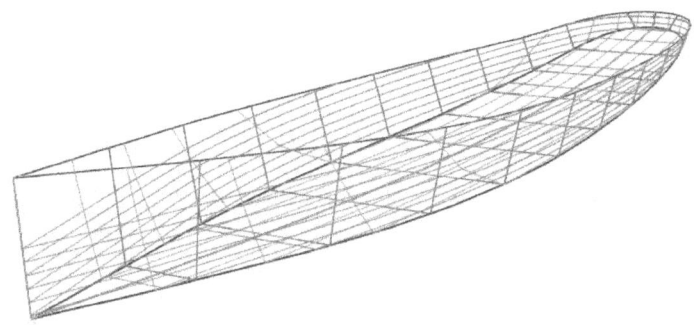

The shape of a New Haven sharpie, which had a low deflection angle and little rocker for most of its length. Illustration by John MacBeath Watkins

In terms of the ratios George O'Day proposed, counting only sail carried to windward the SA/D ratio is 87, counting all downwind sail, 132. (For the more moderate rig I've illustrated above, the sail area/displacement ratio would be 63.5 counting only sail carried upwind.) O'Day said the ratio needed to be 28 or higher. For a sharpie carrying the same weight with a working rig, the ratio would be 26.4. The displacement/length ratio would be 56, based on a sailing length of 33 ft. on a 35' sharpie, which is far better than the 80-125 O'Day suggested. The sails were self-vanging, and the springboards added plenty of sail-carrying power. About the only thing working against them was the amount of rocker aft, which was necessary for them to carry a full load of oysters when they weren't racing.

How fast were they? From Howard Chapelle's *Migrations of an American Boat Type*:

"A large sharpie was reported to have run 11 nautical miles in 34 minutes, and a big sharpie schooner is said to have averaged 16 knots in 3 consecutive hours of sailing. Tonging sharpies with racing rigs were said to have sailed in smooth water at speeds of 15 and 16 knots."

The hull speed of a 35-foot sharpie would be about 8 knots, and planing speed would be anything above 14 knots. These racing sharpies were at least semi-planing, and most likely fully planing boats.

But racing in such boats didn't last long. As engines became available, the sort of oyster fisherman who had the leisure and money to build a racing rig would have been the first to move to the new technology, and unlike log canoes held together with wooden pegs, a sharpie fastened with steel nails didn't last long. Nor do I know of anyone who has built a replica of a large racing sharpie.

The next group to take up the banner of skimming across the water at high speed was the sailing canoes. In the mid-1880s, sportsmen were racing canoes in a serious manner, because it was about the cheapest racing sailboat you could get. I've seen film footage from 1915 of an open canoe planing, and the decked canoes were faster. Paul Butler, who didn't weigh much, added a sliding seat to his canoe, *Vesper*, built in 1886.

Vesper was a very flat-floored canoe, as you can see.

As it happens, in 1886 there was an international match between some British canoes shipped to New York and the American and Canadian canoes. The British had pursued heavier, ballasted canoes. In addition to *Vesper*, they faced other canoes that were light and flat-floored, unballasted and kept upright entirely by an active crew.

They took what they learned back to England, where they built boats that were designed to plane.

The day of the heavy displacement English canoes Pearl and Nautilus ended in 1886, when the two latest models of these renowned types visited the United States, Mr. Baden-Powell bringing his *Nautilus* and Mr. Walter Stewart a new *Pearl* - both sailed from below deck and heavily ballasted. After being decisively defeated by the light American canoes *Vesper* and *Pecowsic* at the meet at Grindstone

Island, they were beaten by the New York C. C. boats on New York Bay, though their owners abandoned the below-deck position and sailed them from the deck. This experience led to the introduction of light displacement canoes with many American fittings in England, and also resulted in the production of a new type of sharpie canoe, with flat floor and straight sides, sailed without ballast. For some years following 1890 canoeing and canoe racing were in a very depressed state in England

In addition to *Shadow*, which seems to have come out in 1891, there are lines available for the *Isalo*, sharpie canoe of 1891:

and the *Isalo*, sharpie canoe of 1891:

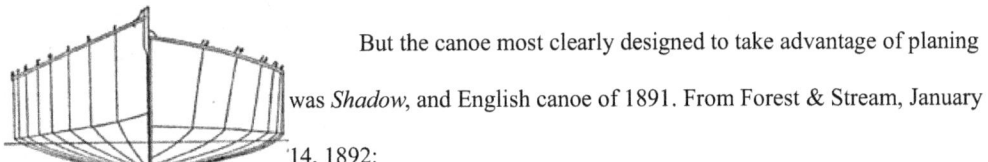

But the canoe most clearly designed to take advantage of planing was *Shadow*, and English canoe of 1891. From Forest & Stream, January 14, 1892:

The accompanying drawings, to which we are indebted to the *Model Yachtsman and Canoeist*, show a peculiar type of racing boat that has recently come into use in England under the title "canoe-yawl," though very different from the boats commonly classed under that elastic and comprehensive title. While their birthplace was on the Thames, they are obviously American in type, and unlike any of the native small craft. We quote the following description by the builder:

The *Shadow* is not, as some may suppose, the result of a "happy hit" in the way of design, but is rather the result of careful; original thought, based upon close observation of the performance of various types of boats of light displacement that have appeared on the river at Oxford. Although the first of the Oxford canoe-yawls she was preceded by several boats of the sharpie type, which were purely experimental, the first

of these being the *Yankee*, followed by the cat rigged *Domino*, the sloops *Merlin* and

Skipjack, and the canoe *Iris*, boats which have in turn under favorable circumstances

shown a remarkable pace. For instance, the *Domino* might have been seen careering

over Port Meadows with about 12in. of water under her at a pace that could not be short

of 10 to 15 miles an hour. This occurred three years ago, during a strong S.W. wind; and

instances have been noted when the sharpies have gone apparently three times the pace

of other boats in competition,. By a peculiar adjustment of the surplus buoyancy and the

displacement of the Oxford yawls have the faculty to a greater or less degree of

"skidding" over the water, and not "wallowing" in it as most boats do. The same faculty

has been attained even in the round-bodied boats, such as *Wisp* and *Torpedo*.

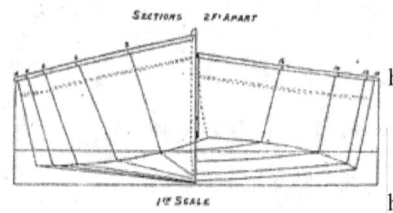

Shadow was 18 ft. by 4.5 ft. With her powerful, flat

hull, she would certainly excel in reaching in a brisk wind.

But the English canoe scene shifted more toward

heavy, cruising canoe yawls.

Of course, there were other planing types. It was well known that scows, garveys, and

sneakboxes were fast reaching in a fresh breeze, but the first two were not too good in light air, which

is the prevailing condition in many American venues. In fact, after Americans apparently developed the

first planing sailing canoes, the class came to be dominated on this side of the water by canoes with

rounded sections, intended to be at their best in light winds. It was not until Uffa Fox and a friend came

over with his reinvention of the planing canoe that Americans began building planing sailing canoes

again.

But after the sailing canoe racers abandoned the planing form, the idea of going faster than the competition in the right conditions still had a strong appeal. In the winter of 1892-1893, a boatbuilder named Arthur Dyer built a 24-foot sloop in a shop with the windows painted black so that no one could see in. The boat was *Onawa*, a shallow, flattish boat with a long spoon bow and a broad, flat counter. In those respects, she resembled *Alpha*, a Nathaniel Herreshoff design

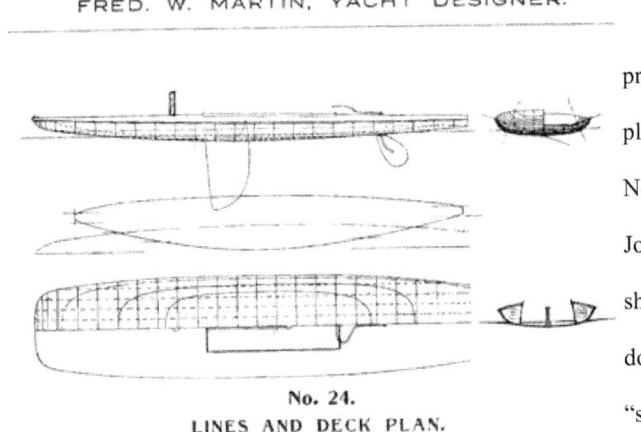

An 1880 drawing of an American sneakbox by Dixon Kemp

of 1891, but she was lighter and seems to have had much less rocker.

Both of these boats owe a debt to a smaller, older type of boat, the sneakbox. Originally developed for hunters who wanted a low-profile, shallow boat, they had an arc bottom and a flat run.

FRED. W. MARTIN, YACHT DESIGNER.

Midwesterners played a prominent role in developing planing boats. In 1896, a young Norwegian immigrant named J.O. Johnson designed a square-ended, shallow boat 38 ft. long, which the dockside admirals derided as a "scow." The boat proved far faster

No. 24.
LINES AND DECK PLAN.

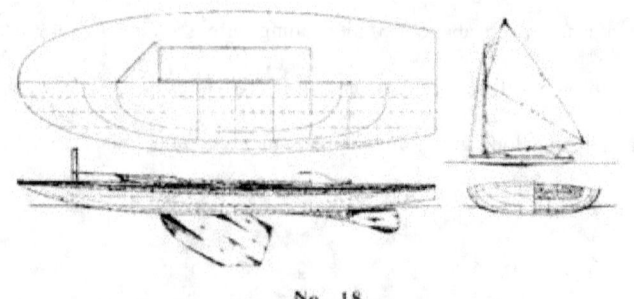

No. 18.

LINES, DECK AND SAIL PLAN.

than any of her opponents, and inland scows were soon recognized as some of the quickest boats in the world.

By 1901, another midwesterner, Fred Miller, was advertising boats he sold as knock-down kits, including some pretty modern looking scows and some catboats that appear to be based on sneakboxes

Rules were promulgated, and soon there were various classes of scows, including the Class A scows that are the size of Johnson's original design.

By the time Miller put out his 1901 catalog, the scows looked much like they do today, at least in hull form.

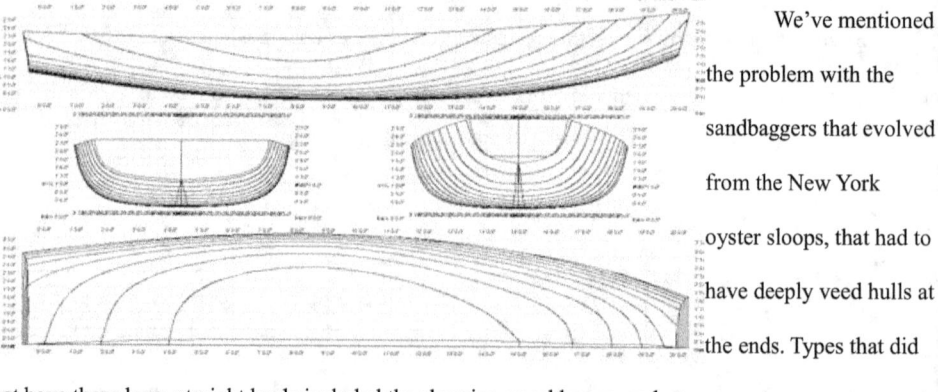

We've mentioned the problem with the sandbaggers that evolved from the New York oyster sloops, that had to have deeply veed hulls at the ends. Types that did not have these long, straight keels included the sharpies, sneakboxes, and prams.

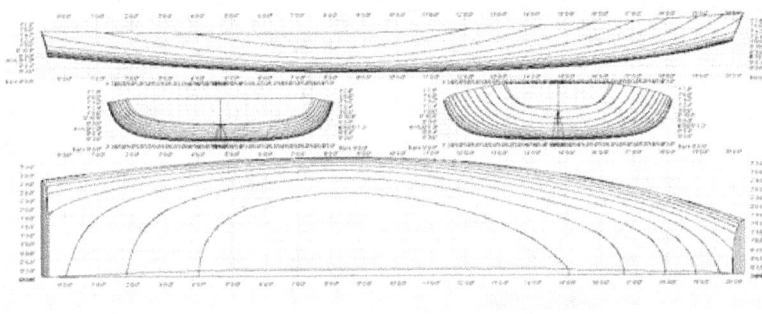

Consider the traditional shape of the pram, and consider what happens when you take those lines, and reduce the vertical scale, squashing the shape down: A reduction to 60% of the vertical scale turns a pretty ordinary looking punt into something that looks a bit like the boat J.O. Johnson's friends derided as a "scow."

Once the idea that flattish boats were the quickest in certain conditions became established, news spread quickly. In both Britain and in America, the half-rater class caught on, and by the time an 1896 article in *Outing* magazine was published, about a hundred had been built in America. Some had what clearly looks like planing hulls.

This gives a feel for how flat the boat were in cross-section, a picture of the Clinton Crane design, *El Heirie*. Some of these boats, by the way, were built with an inner layer of 1/8-inch planking and an outer layer of ¼ inch planking with a layer of silk between, and cost about $1,200. You could probably have bought a modest house for that.

One of the top boats cost only $300. That boat was *Question*, an arc-bottomed sharpie built by Larry Huntington.. It seems to have been shaped like a lighter, flatter version of the Star class keelboat. I've been unable to find lines for this boat, but here's a photograph that ran in Outing magazine in August, 1896:

Of course, sharpies were known to be fast long before this, and before the racing sharpies of the 1880s. In 1855, the redoubtable Bob Fish built an experimental sharpie with a shallow V hull named

Luckey.

The vessel was 54' 6" long and 15' 9" wide, and certainly not a planing boat, but I

El. HEIRIE

doubt Fish would have built a boat of this size and shape if it had not proved successful on a smaller boat. The problem is, until sportsmen took an interest in having official races for small boats, their lines were seldom recorded.

But by the 1890s, several publications were printing plans and photographs of fast small boats Part of this is a new focus for sportsmen, but part is new technology, as photography and photolithography progressed. This spread knowledge more quickly.

In 1852, knowledge of the American type of boat spread to Great Britain when Una and a 20' sloop named Truant were shipped to England and impressed people with their speed. Truant, in particular, seems to have been taken around the country and won races. Later in the 1850s, three Britsih

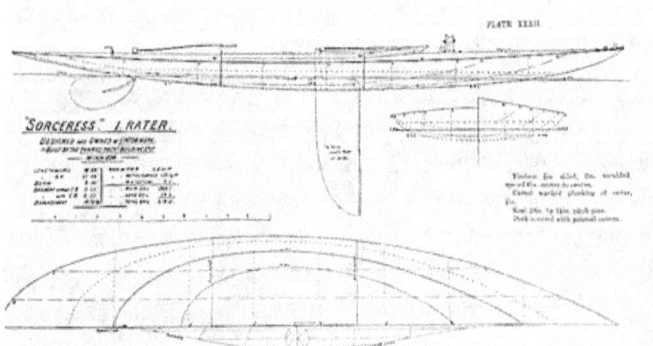

centerboard boats were shipped to Australia, *Charm, Presto* and *Challenge,* which impressed Australian sailors with their speed.

By the 1890s, the way knowledge was

transmitted had changed. Only a year after Dyer launched *Onawa*, Linton Hope, in England, designed a Thames 1 rater that was similar except it lacked the wide, flat stern that helped *Onawa* plane.

In New Zealand, James Clare built a boat named *Maka Maili* in 1898 influenced by American designs, but the design seems to have been his own. The same year, Logan Brothers, a firm of Scottish immigrants, designed and built a restricted half-rater class (essentially a one-design that could race with other half-raters in the development class as well as in its one-design class) with flattish sections called the Patiki Class. Scow-like New Zealand racing sailboats became known as Patikis, and they thrived in the protected harbor of Hawke's Bay at Napier, on the east side of the north island of New Zealand, where they had good winds and protection from big seas. When the 1931 Napier Earthquake caused part of the harbor to collapse into the sea, they were exposed to rough seas that tended to break the hulls apart.

This points to a major flaw in the early planing designs. They needed fresh winds to take advantage of their ability to plane, but they didn't do well in rough water, so they tended to be best suited to lakes and large, well-protected harbors like Barnegat Bay, where the sneakboxes evolved into something like scows.

In much of the United States, either the winds favored boats with a distinctly light-air orientation, or the waves were a problem for scow-like boats. Even sharp-bowed, flattish boats like the 1908 SS Class required someone bailing most of the time going to windward.

In New Zealand, the X class of 1921 was strong enough to take the waves, but their rather bluff bows must have slowed them down going to windward in a seaway. In any case, they were an early planing type. The class was proposed in 1916, with a set of lines drawn up, although it was an

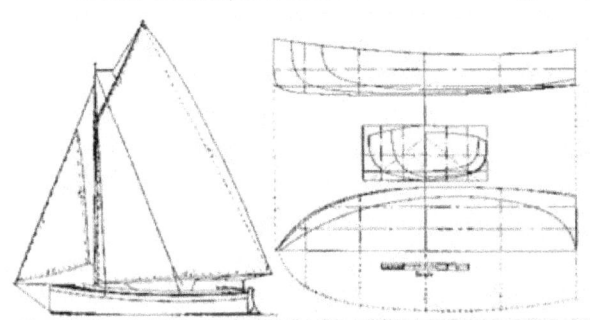

The 1916 drawings of the 14 ft. one design class.

open class.

The problem remained, how does one take advantage of the speed potential of planing while still having a boat that is fast to windward in a seaway, and doesn't get beat up too much from pounding?

The solution came from England, where Uffa Fox, who had apprenticed working on flying boats and powerboats, applied what he had learned to sailboats.

At the time, some of the fastest powerboats were "autoboats," a type that tended to be deeply veed forward and flat aft. The deep V forward meant a greater deflection angle to the hull, throwing water to the sides instead of raising the boat up, but Fox felt he could get by that by making the boat flat aft and shifting weight aft when the boat was reaching at speed. Because the resulting shape was deep chested, you could keep the transom out of the water by moving your weight forward when you weren't planing.

The veed shape meant these boats would increase their resistance quickly as they heeled, unlike

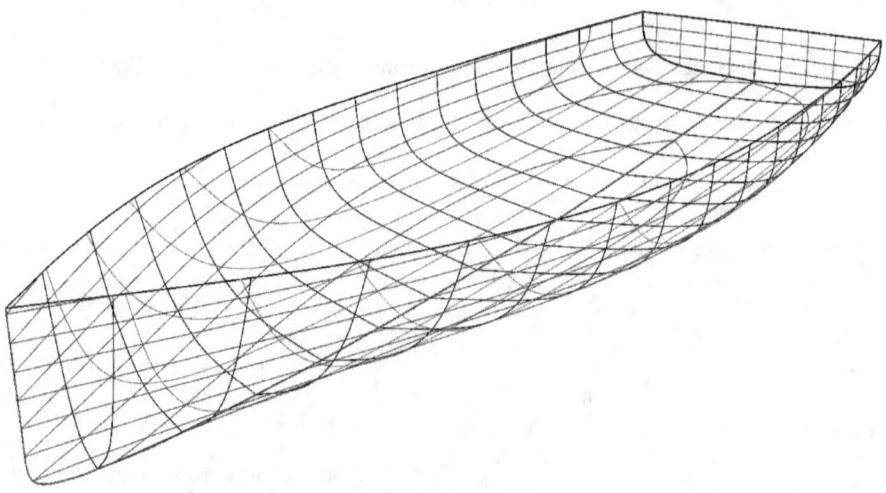

Th
e type of planing dinghy developed by Uffa Fox. This is not a particular Fox design, just my impression of how these boats are shaped in general. Illustration by John MacBeath Watkins

the flat-bottomed scows that had to be heeled going to windward. I've drawn up a set of lines that will illustrate the Fox type of planing hull (It's not meant to represent a particular boat.) You will note that unlike the American type, which is flatter forward, it does not need much rocker at the stern to keep from dragging the transom. When I first saw an Uffa Fox hull (a Firefly) it struck me as a displacement hull that couldn't possibly plane, because of its deep V at the stagnation point. An English friend responded to boats like the C Scow and the Lightning by thinking they couldn't plane because of their rocker aft.

The boats Uffa Fox designed could do what the scows and sneakboxes could not, it could go to windward in exposed waters and a steep chop and move as well as boats with rounder sections, while planing on the reaches. This meant that they could be used in any venue without slamming down on waves and breaking the hull. The boats with rounder forms, such as the designs of Morgan Giles, could plane in optimum conditions, but Fox's boats from his 1928 *Avenger* on could plane in the conditions the National (later International) 14 class usually encountered on a typically windy English summer day.

This led to a new generation of planing boats which began to spread through the world. Australian 18-foot skiffs had concentrated on carrying massive rigs and heavy crews, producing what amounted to a heavy-air version of the sandbaggers, but in 1932, *Aberdare,* with a smaller crew than usual, lighter weight, and a hull designed by a collaboration of her owner, Frederick Hart, and her builder, H.P. Whereats, introduced planing to what is now the fastest planing class in the world.

Modern planing hulls tend to

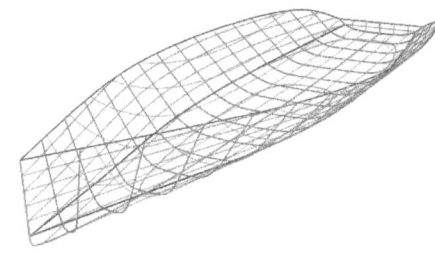

Modern planing dinghies tend to have narrow entries, U-shaped sections forward, flattening out to the stern, and very little initial stability, with flare or hiking racks to give the crew more leverage, and often trapezes as well. Illustration by John MacBeath Watkins

be lighter, and can take advantage of the flat forward sections of the American planing type along with the straight run of the English type. A modern planing hull would look something like this design I drew up to illustrate the type:

Of course, getting the hull shape right and making the boat light enough are important, but there's more to getting boats to plane than that. You have to put a powerful rig in the boat, and you have to be able to use that power, which means you need power to carry sail. Some modern planing boats do this with hull flare, like the Merlin Rocket. Some use hiking racks or trapezes.

You also needed sails that were powerful on a reach, which is where most planing happens. If the sail twists too much on this point of sail, part of it will be stalled or part of it will be luffing, and twist happens when the boom rises.

On the racing sharpies we mentioned earlier, the power to carry sail came from a flat-bottomed, shallow hull, two long springboards, and a crew of a dozen active men. The rig on these boats was also self-vanging. Because the sprit booms go across the sail, the foot of the sail is under tension any time the boom tries to rise. A trick Nat Herreshoff used on many of his boats was to have a light stick to tie to the mast and the boom, so that it was under compression any time the boom tried to rise. The modern vang seems to have been invented in the 1930s or perhaps the late 1920s in the International 14 class

And of course, boats with a spinnaker tend to plane sooner than boats without one. The evolution of downwind kites is a bit beyond the purview of this essay, but it's worth mentioning that the single-luffed spinnaker, carried entirely on one side of the forestay, was replaced by the double-luff parachute spinnaker in 1927, the new sail first appearing on the 6-meter *Maybe*. As boats got faster and skippers got more sophisticated about tacking down wind, a new single-luffed spinnaker, the asymmetric now carried by International 14s and other fast classes, replaced the parachute spinnaker on many classes.

The power to carry sail without tons of ballast is at least equally important. By 1874, when Thomas Eakins painted *Sailboats Racing on the Delaware*, "hiking" (as in, taking a hike to the high side of the boat) was common. Ben Fuller, curator of Penobscot Maritime Museum, has documented the method used to put most of a crew member's body weight well outside the boat on the weather side. He says they used "a short length of line clipped into a ring on keel with a T handle on it. You tuck your feet under the line and hang onto the toggle."

That sounds as effective as modern hiking straps in getting the crew weight where it does the most good, but would have taken one hand of the person hiking to keep a grip on the handle, so it's not as good for racing on the whole as modern straps, apparently developed in the International 14 class.

A close examination of the 1874 picture of the crew racing a hiker shows that the man at the helm has his weight inboard. That's because he doesn't have a tiller extension. When the sailing canoes developed sliding seats in the 1880s, this was followed by a number of ingenious inventions to give the skipper a way to control the rudder. In *Breezing Up,* a Winslow Homer painting from about the same time (in a boat that looks a bit like a Woods Hole spritsail boat), the helmsman is sitting well to windward, controlling the tiller with what looks like a piece of line, which works well enough if you've got plenty of weather helm.

Here Uffa Fox appears again, bringing a better planing hull from England to race against the American canoes, and a modern tiller extension, which seems, like so many things, to have been invented in the International 14 class.

Thames raters experimented with a hiking system developed by Beecher Moore, owner of *Vagabond*, called "bell ropes," where the crew grabbed a line attached to the mast, stood with their feet on the top of the gunwale, and put their entire weight overboard. Something similar was used by Malay sailors racing kolec canoes in the 19th and early 20th centuries. H. Warrington Smyth wrote about them first in a 1902 journal article, later in his 1905 book, *Mast & Sail in Europe and Asia*. They stood on the gunwale of canoes that ran to about 45 feet, and swung their weight outboard while holding man-ropes.

But again, it was the International 14 class that invented the modern trapeze, with the crew wearing a harness with a hook on it that the trapeze can clip to, leaving the crew's hands free while they put their feet on the side of the boat and their entire body weight over the side.

The Delaware hikers apparently used hiking racks, but now many boats that use them also have trapezes, allowing the crew to place their weight even farther from the centerline and again increasing the power-weight ratio possible on a sailboat.

In the Moth class, which specifies 11 feet in length and 80 square feet of sail, the scow and skiff types met on even terms. The class was dominated in Australia for years by scow Moths, except in light winds, but before the shift to foiler Moths, the skiff type developed higher speeds, even down wind. The scow Moths had evolved into flyweights made with 1/16-inch plywood and plenty of light internal bracing, and could recoup ground lost on the windward leg by planing more readily on the downwind legs. The skiff type, with their pointed bows, tended to be faster to windward in a chop as well as in light winds, making them better for racing in areas like Florida, and as they developed a modern shape with a narrow waterline and shallow forefoot, they became about as fast down wind as the scows.

A further evolution of the planing type is boats that use the rig not only to drive the boat, but to help lift it. After multihulls pretty well dominated the sailing speed record in the 1970s, after 1980 most of the records for maintaining a high speed over short distances were set first by windsurfers, which are a type of planing sailboat, then by kiteboards, which don't even have enough buoyancy to stay on top of the water with the sailor until the kite starts to pull. These have a minimum of weight and wetted surface. The current record holder is the *Vesta Sailrocket 2*, which rides on three stepped planing hulls, foils, and has a canted wing sail.

The increasing division between daysailers and racing skiffs

The evolution of planing sailboats has favored the fast over the pleasant, at least in development classes.

Here, for example, is an interpretation of current trends in planing sailboats:

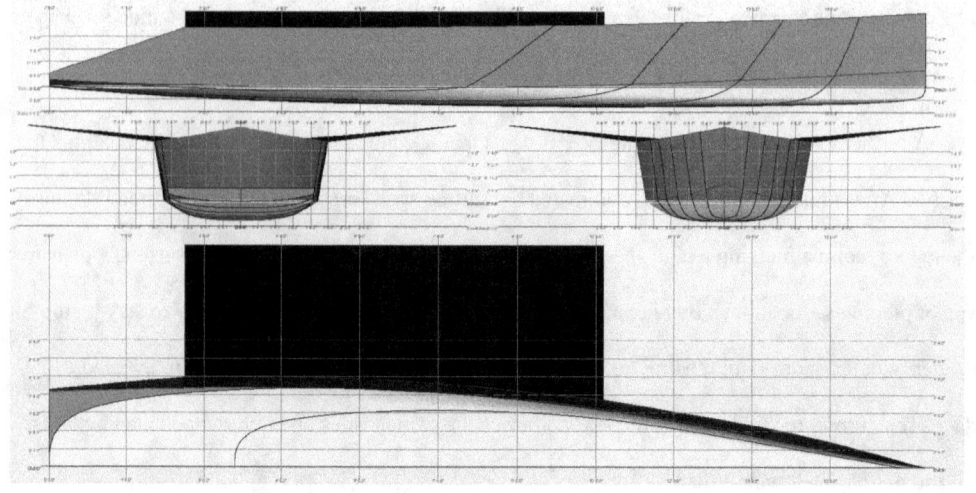

Hull A. Illustration by John MacBeath Watkins

This represents a 17-foot racing dinghy with minimal initial stability, a high prismatic coefficient (.62) and hiking wings. The waterline beam is about 3 feet, the beam of the hull is about 3½ feet, and the beam of the hiking wings is about 8 feet. The entrance angle is about 9 degrees. It would carry about 200 square feet of sail, have a crew of two, both using trapezes to increase their righting moment, and it would have a reefing bowsprit for an asymmetric spinnaker. Everything, the hull, the rig, even the hiking racks, would be built with carbon fiber so that its scant displacement can support two highly skilled sailors.

And they'd better be highly skilled, because although fast, the boat will be nearly impossible to sail. It is designed to have the least possible resistance going to windward in waves, thus the narrow hull. The theory is that modern planing sailboats are fastest sailed flat, so there's no point in making them stable, because the crew will sail them so flat the center of buoyancy has no opportunity to move to leeward.

The type can be used without the trapezes or the jib and spinnaker. In fact, it started with the Moth class. The last of the non-foiling Moths looked quite like this, but were 11 feet long with a waterline beam of about one foot. I've heard of an Australian Moth champion starting 5 minutes behind the legendary Flying Dutchman class and finishing the race having overtaken all but the leaders of that class. The down side is that almost no one can sail a boat like this. It looks like the central hull of a trimaran, and has about as much stability. Happily, even most development classes use wider hulls, in part because of class rules controlling beam.

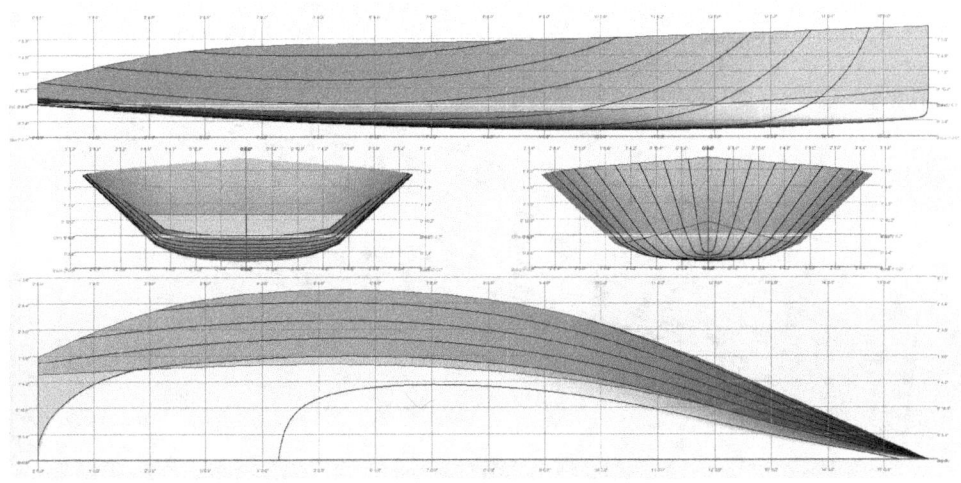

Hull B. Illustration by John MacBeath Watkins

There is a more moderate approach in hiking classes that do not allow hiking racks. Consider this boat, about 16 feet long, 3 1/2 feet wide on the waterline, and just under 6 feet wide at the deck, which we will call Hull B. I initially drew it as an exercise in what a boat would look like if you designed a modern hull for the Snipe class rig.

This is quite moderate compared to a modern Merlin Rocket, which would have a beam at the deck of about 7 feet and a waterline beam of about 3 feet. The entrance angle is a little over 10 degrees, and it has the same U-shaped sections forward as its narrower cousin, flattening aft to help it plane. The flare allows the boat to gain some stability before it reaches the point where capsize is inevitable, but this is still based on the idea that stability should be minimized to reduce the drag of the hull. It's just the same set of priorities applied to a boat that more people could successfully sail around a race course.

Compare this to the older planing types.

The first type to plane regularly during races seems to have been the sharpie, as rigged for racing in the 1870s:

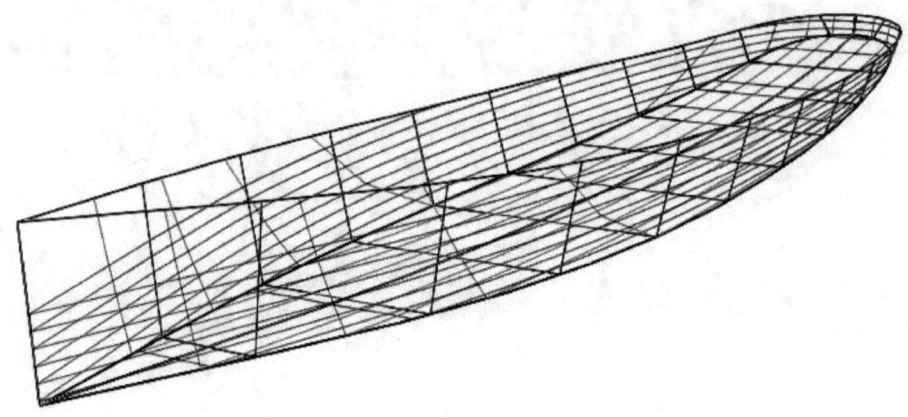

Sharpie hull. Illustration by John MacBeath Watkins

These flat-bottomed workboats, with the moderate rig usually carried, had a useful amount of stability for getting to and from the oyster beds, and for tonging for the oysters while there. With the truly frightening rigs they carried on race days, they carried ten men on springboards to stabilize them, and more than 1,000 square feet of sail to windward, more reaching and running. The boat would have been about 35 feet long and 7 feet wide, and the unballasted hull would have been light, probably less than 3,000 lb.

The second type known to have regularly achieved planing speeds when racing under sail were scows, with a flattish arc bottom and a bluff bow, relying on the small angle of between the rise of the bottom to the bow for their entry angle:

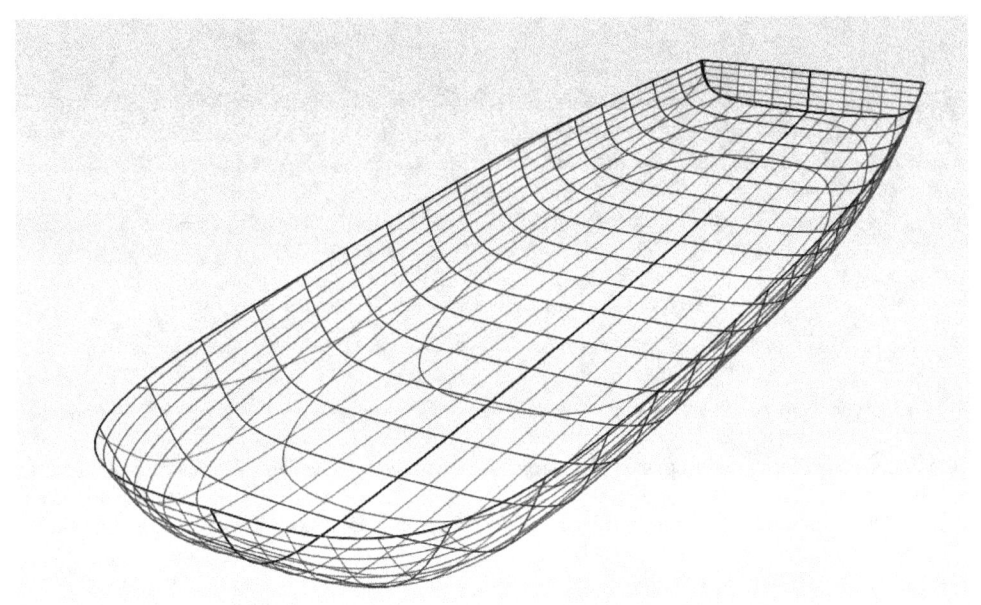

Scow hull. Illustration by John MacBeath Watkins

These boats have very stable hulls that extend the maximum beam farther forward than a sharp-bowed skiff hull. Scows are sailed heeled (to leeward) when going to windward, and fairly flat reaching in a strong breeze. As a result, they are useful daysailers, which can be sailed in a relaxed manner when not racing. When racing, the hulls have a minimal deflection angle at the forward part of the hull, so the crew doesn't have to move aft when the boat is planing. This is true of both the scows and sharpies, even more modern sharpies like the Lightning class. Now consider the type Uffa Fox introduced with *Avenger* in 1928:

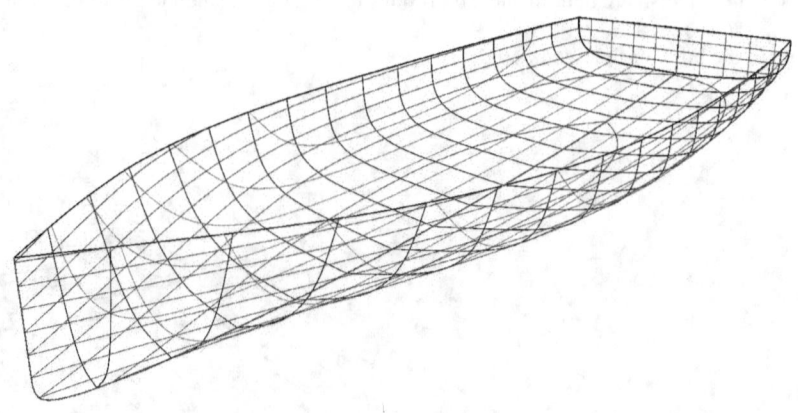

Uffa Fox-type hull. Illustration by John MacBeath Watkins

This type is deeply Veed forward. This allowed these boats to be fast in choppy conditions that hit the flattish scow bows and impaired their progress to windward. The deep Vee forward produces too great a deflection angle for planing, which means that in planing conditions, the crew must move aft. The type was more stable than the U-sectioned boats that preceded it in the International 14 class and similar classes. To sail it fast, you needed to sail it flat, but the inherent stability of the type made it forgiving. They are fastest sailed flat, but if you failed to sail them flat, they simply became less efficient. You could capsize them, but not as easily as the modern types.

The same year *Avenger* came out in England, Ted Geary designed a planing dinghy for the Seattle Yacht Club. That design, originally called the Flattie and now called the Geary 18, had a sharp bow and sharpie-type construction, but the bow was well clear of the waterline and the hull was clearly a scow type. The class added more sail area and a trapeze as time went on, but no spinnaker. They have a faster Portsmouth Yardstick rating than the Jet 14, which combines a 1950s-era International 14 hull with a Snipe rig, with a spinnaker added, but no trapeze.

All of these older types were planing racing boats that could be used for daysailing. Few classes made this as explicit as the Lightning class, which sent a design brief to Olin Stephens asking for a boat sailors could take their families out in, but also occasionally race.

These are the types of vessel that led to the great age of dinghy sailing from the 1950s through the 1970s. We now see the best racing sailboats as something unsuitable for taking the family sailing in. The older classes are still suitable for use as daysailers and as racing boats. To some extent, the older one-design classes have mitigated the effect of trends toward a division between daysailers and racing boats, but some have allowed costs to skyrocket.

The Optimist Pram, the most popular sailboat in the world, no longer allows homebuilt boats, and allows carbon fiber spars.

Now, I've built a 9' 6" dinghy for less than $500 in materials that will easily out-perform the 7'9" Optimist, but I can't buy an Optimist new for less than $3,000, and can easily spend $4,500 for one with carbon spars and a boat cover.

The problem for one-design classes is that the most avid racers are likely to be the most active participants in running the class, and they will want to be allowed to use the latest go-fast items. It's easy for a class to become as expensive as the latest racing designs without matching their performance.

We are likely to continue to see a division between daysailers and racing boats as time goes on and the older designs are left behind. The classes trying to continue to provide a more stable boat, such as the JY 15, tend to be those which are aimed at youth sailing programs, not at families. There are still companies providing kit boats, but these tend not to be the sort of racer-daysailer that kit boats tended to be in the 1960s. Chesapeake Light Craft, one of the more successful kit providers currently, produces not one racing dinghy class. Not everyone likes to race, but it would be nice if people could experience the handling of a nice planing dinghy without either getting a boat that is difficult to sail, or one that harks back to the 1930s.

Let's see where the logic of this situation takes us. We want a dinghy that is stable enough to that most sailors can handle it. We want enough carrying capacity for a mom, a dad, and a child. Current practice says that it should have a narrow entry, U-shaped sections forward, a flat area aft of the mast and ahead of the transom on which to plane, and enough flare that the crew gets some extra leverage for hiking, but not so much that it's difficult to get on and off it from a dock.

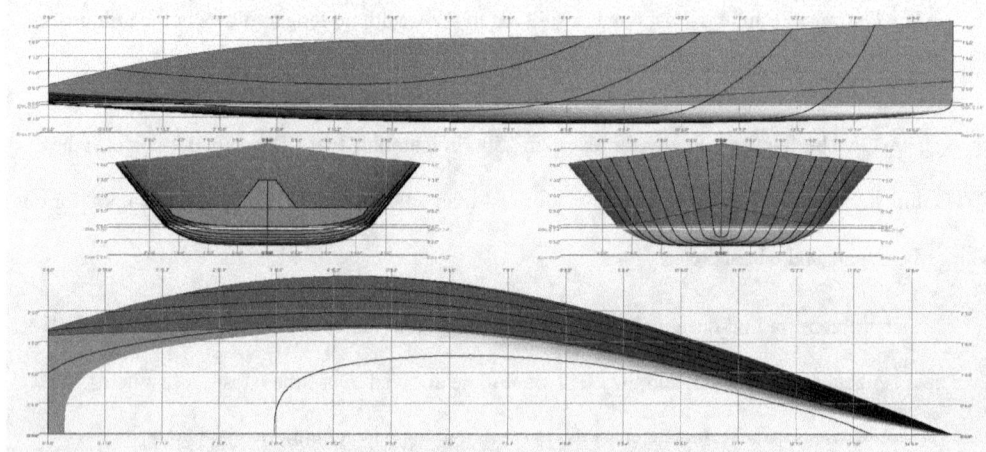

Hull C. Illustration by John MacBeath Watkins

This boat is 15' by 5'. You'd want to build it with a watertight deck above the waterline, so that it could be recovered easily after a capsize, so it's designed to drain out the stern.

Now consider a more forgiving design.

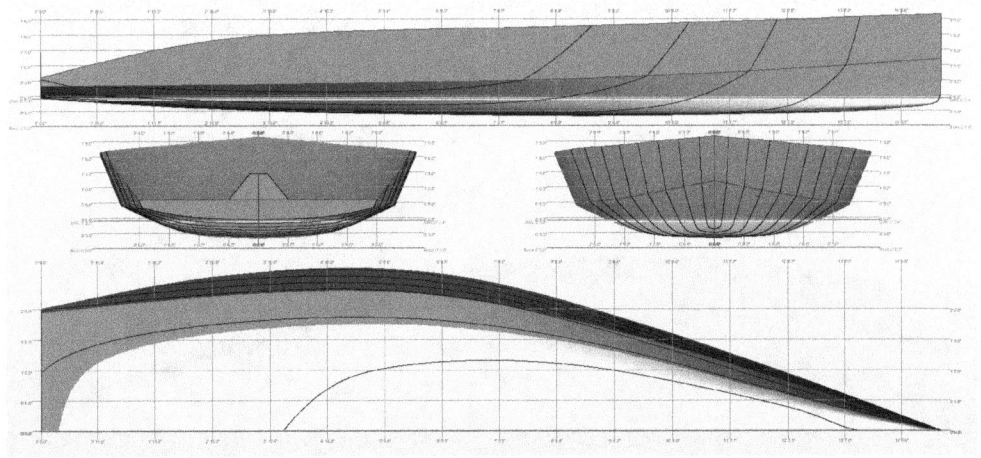

Hull D. Illustration by John MacBeath Watkins

This is a more stable, forgiving boat, because of the chine above the rather narrow waterline. It will plane easily because it has a low deflection angle and straight run. It has a narrow waterline and substantial flare, with chines a few inches above the waterline. Because the flare is close to the waterline, it would be affected by waves when going to windward. In fact, the sort of sailor capable of sailing the narrow boat with hiking racks, it would seem like a racing boat with training wheels. As a boat for training sailors, it would have the advantage of being a boat that is easy to sail, but challenging to sail fast. It would certainly be fast if sailed flat in flat water, but chop or lumpy water would slow it down.

Finally, let's consider an even more stable and forgiving boat:

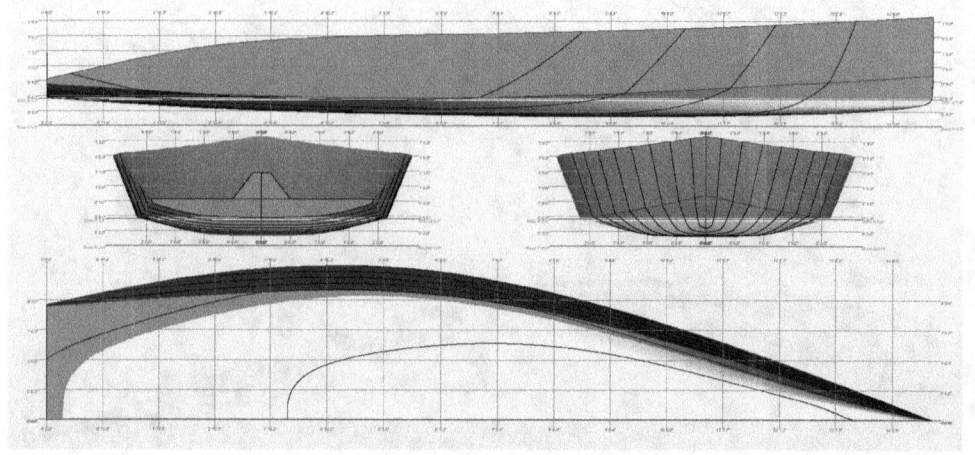

Hull E. Illustration by John MacBeath Watkins

This boat, with a waterline beam of about 4 feet on a length of 15 feet, would be the pleasantest and most stable daysailer of these three. The wider waterline beam would make it slightly slower to weather in a chop than hull C. I've attempted a version of hull B, but without the extremely narrow waterline or the extreme flare to the topsides.

These exercises are intended to give an idea of the sort of compromises needed to produce something that incorporates what we've learned about hull shapes without forgetting what we've learned about what makes a boat a practical daysailer.

Doubtless there are people better qualified than me trying to solve the riddle of making a practical daysailer with good performance, and some of the older designs are still quite popular, for good reasons. Some classes, like the El Toro and the Windmill, still support the home builder. And demographic patterns, such as the trend toward more people living in cities, make it difficult to find space to build or keep a boat of any sort. But the separation between racing and daysailing boats continues apace, and it doesn't seem like the best way to promote the sport.

www.ingramcontent.com/pod-product-compliance
Lightning Source LLC
Chambersburg PA
CBHW071122220526
45467CB00004B/2019